DARWIN IN

In the same series by the same authors:

Curie in 90 minutes
Einstein in 90 minutes
Halley in 90 minutes

forthcoming titles

Faraday in 90 minutes
Galileo in 90 minutes
Mendel in 90 minutes
Newton in 90 minutes

John and Mary Gribbin

DARWIN
(1809–1882)
in 90 minutes

Constable · London

First published in Great Britain 1997
by Constable and Company Limited
3 The Lanchesters, 162 Fulham Palace Road
London W6 9ER

ISBN 0 09 477050 6
Set in Linotype Sabon by
Rowland Phototypesetting Ltd,
Bury St Edmunds, Suffolk
Printed in Great Britain by
St Edmundsbury Press Ltd,
Bury St Edmunds, Suffolk

A CIP catalogue record for this book
is available from the British Library

Contents

Darwin in context

Charles Darwin did not invent the idea of evolution, which had grown steadily from the sixteenth century onwards as scientific learning developed in Europe. His important contribution was to offer an explanation of how evolution works – natural selection. Even here, though, he was not unique, as we shall see. Darwin's rare value was in the breadth of his learning; the way in which he applied the scientific method to all his work, leaving no room for wishful thinking; and the clarity and beauty of his writing, which got his message across (and still does) to a wide audience. More than for any other great scientific advance, it is still true to say that the best way to learn about evolution is to read the original works of the scientist who put the idea forward.

What you might call evolutionary ideas (though rather strange ones) about how the living world got to be the way it is can be traced to the time of the ancient Greeks. We pick up the story with Francis Bacon (1561–1626), who, in his book *Novum organum* published in 1620, wrote about the way in which species vary naturally from one generation to the next. He pointed out that such natural variation could be used by the breeders of plants and animals to produce 'many rare and unusual results'. In Germany, Gottfried Wilhelm Leibniz (1646–1716), best remembered as a mathematician, was intrigued by fossils and by the possible relationship between the extinct ammonites and living species such as the nautilus. Speculating that the species had changed because the modern variety lived in a different environment from their ancestors (he did not actually use the word 'evolved'; the term 'evolution' was first used in its modern biological context in 1826, by Robert Jameson), he wrote, 'it is credible that by means of such great changes [of habitat] even the species of animals are often changed'.

In the seventeenth century, this was by no means an isolated view. In the eighteenth, Georges Louis Leclerc, Comte de Buffon (1707–88), puzzled over the geographical distribution of similar but different species, and suggested that the North American bison might be descended from an ancestral variety of ox that had migrated there, where 'they received the impressions of the climate and in time became bisons'. Such ideas implicitly include not only the concept of evolution, but also the notion that species have evolved to fit their environments – adaptation.

The greatest of Darwin's predecessors as an evolutionary thinker, though, was his grandfather Erasmus Darwin (1731–1802). A larger-than-life figure, he was a doctor good enough to be asked by George III to become his personal physician (though he did not take up the post), a poet good enough to be considered a candidate for Poet Laureate (though he just missed out there) and a 'natural philosopher' good enough to be elected a Fellow of the Royal Society (though his most important piece of scientific work now bears someone else's name). Erasmus combined several of his interests by writing about evolution in a long erotic poem, *The Botanic Garden*, as well as in a two-volume prose work, *Zoonomia*, published in 1794 and 1796, more than ten years before Charles Darwin was born.

Erasmus Darwin was clear about the importance of evolution, but mistakenly thought that individual members of a species developed different characteristics during their lifetime, and these enhanced characteristics were passed on to their offspring. In his picture, for example, if a deer needs to run away from a predator to avoid being killed, it will somehow acquire a body slightly better adapted to running – more or less by willpower – and pass that slightly improved body plan on to the next generation through heredity. The same process, repeated over very many generations, then 'explains' why deer are such good runners today.

Curiously, almost exactly the same idea was developed independently in France by Jean-Baptiste Lamarck (1744–1829), who published it in 1801. Lamarck was a full-time scientist, Professor of Zoology at the Museum of Natural History in Paris, and developed the idea much further, so it is perhaps just that it should be known today as 'Lamarckism' (it would certainly be confusing if it were called Darwinism!). What is less just is that it is often derided as a silly idea, when in fact it was for its time a respectable attempt to put evolution on a scientific footing, and helped to stimulate discussion about the concept of evolution.

Lamarck's ideas were taken up by his discipline Étienne Geoffroy Saint-Hilaire (1772–1844), who not only promoted them but made the first clear statement of what we would now call the survival of the fittest. Writing in the 1820s about the kind of modifications to an individual that Lamarck (and Erasmus Darwin) had described, he said:

> If these modifications lead to injurious effects, the animals which exhibit them perish and are replaced by others of a somewhat different form, a form changed so as to be adapted to the new environment.

Unfortunately, Geoffroy Saint-Hilaire also had some distinctly barmy ideas about the relationships between species: among other things, he tried to prove that the body plan of a mollusc is the same as that of a vertebrate. This led to all his ideas being judged out of court, and by 1830 Lamarckism, tarred with the brush of Geoffroy Saint-Hilaire, was no longer seen as respectable science in France. In England, hardly anyone knew of Erasmus Darwin's thoughts on evolution. But the young Charles Darwin was on the brink of his greatest adventure. It would be almost thirty years before the fruits of that adventure were presented to an astonished public. But though the public was astonished when the *Origin of Species* was published in 1859, the scientific world ought not to have been. Indeed, had it not been for Geoffroy Saint-Hilaire's excesses, the modern version of the theory of evolution by natural selection might well have been developed from Lamarck's work in France, while Charles was still at sea on board the *Beagle*.

Life and work

We have already introduced one of Charles Darwin's grandparents, so we may as well flesh out his family background by introducing the others. Erasmus, as we have seen, was a successful country doctor. He lived near Shrewsbury, not far from the heartland of the Industrial Revolution in England. He was born in 1731, and became a gentleman amateur in science and the arts, wealthy and influential in science, literature and society. He was also a huge man physically, with large appetites, addicted to life's pleasures – including food, wine and women. His first wife, Mary, died of drink in 1770 at the age of 30, having produced five children; the third of these, Robert Waring Darwin (born in 1766), would become the father of Charles.

For the next ten years Erasmus lived the high life (during this time he fathered two children by the governess of his large brood), before falling, at the age of 49, for a woman sixteen years his junior who was not only one of his patients but married to a rich man. Her husband died in 1780, Erasmus married the widow Elizabeth, and together they shared the parental responsibilities for his five children by Mary, his two illegitimate offspring, and another bequeathed by her late husband. Together the couple produced seven more children before Erasmus, no doubt exhausted, shuffled off this mortal coil in 1802. Elizabeth outlasted him by thirty years.

One of Erasmus's great friends was Josiah Wedgwood, one year older than Erasmus, founder of the eponymous pottery business. One of the first people to make a fortune out of the Industrial Revolution, Wedgwood married one of his cousins, Sarah, then 30, in 1764. The marriage was happy, and decidedly more conventional than Erasmus's second, but no less prolific: they produced three daughters and four sons. Their eldest daughter, Susanna, married Erasmus's son, Robert, in 1796, a year after Josiah Wedgwood had died. Sarah survived until 1815.

The marriage, like much of Robert's life, was largely organized by old Erasmus. Eager to have a son succeed him in the medical profession, Erasmus had been heartbroken when Robert's older brother, Charles, died from septicaemia contracted while carrying out a post-mortem as a medical student. The next son, also called Erasmus, was only a year younger than Charles and already committed to a career in law. Robert, only 12 when Charles died, was young enough to be moulded in the way Erasmus senior wanted, and to become a doctor whether he wanted to or not. Pulling strings and providing financial back-up, Erasmus saw Robert installed in medical practice near Shrewsbury at the age of 20, in 1786. Like his father, Robert was a tall man, tending to corpulence as he grew older. He married when he was 30; Susanna was a year older than him. Robert bought land on the outskirts of Shrewsbury and built a new house, named The Mount, completed in 1800.

The Darwins' daughters Marianne, Caroline and Susan were born in 1798, 1800 and 1803; their first son, Erasmus, in 1804; Charles Robert Darwin was born on 12 February 1809; and the baby of the family, Emily Catherine (known as Catty), arrived in 1810, when Susanna was 44. As Darwin's biographer Janet Browne has pointed out, in the early years of the nineteenth century the life of the Darwin family, and indeed the Darwin family itself, could have been lifted straight from the pages of a Jane Austen novel. Already wealthy thanks to his inheritance (and that of Susanna), Robert Darwin made more money through shrewd investments, particularly favouring property and mortgages on land, but soon moving into canals. At his death in 1848 he left £223,759 – an enormous sum in those days.

Josiah Wedgwood's son and heir, Josiah Wedgwood II, bought an estate 30 miles from Shrewsbury, a country retreat called Maer Hall. The strong links between the Darwins and the Wedgwoods continued into succeeding generations, with several intermarriages. Josiah II himself married his cousin Elizabeth in 1792, and they had nine children. Their first son, Josiah III, married Charles Darwin's sister Caroline in 1837; the youngest of their brood, Emma, was born in 1808, when Elizabeth was 44 years old, as Susanna Darwin was when her youngest child was born.

Charles Darwin's idyllic home life, spoiled by older sisters in a large country house, came to an end in July 1817 when his mother died. Marianne and Caroline were old enough to take on her role in charge of the household, but this left them with less time to indulge their younger siblings. Worse, Robert Darwin was so depressed by the loss of his wife that he forbad anyone to mention it, and his gloom pervaded the family.

Charles (who, as a boy, was actually known to his family as 'Bobby') had started school not long before his mother died, having previously been educated at home by his sister Caroline. In 1818 he was sent as a boarder to nearby Shrewsbury School, which he hated. It was only a mile from his home, but he lived in the school, though he got away as often as possible to spend an evening at The Mount. The best thing about the school (though Darwin made many friends among the boys there) was that his older brother Erasmus (known as 'Eras', or 'Ras') was a senior pupil and could keep an eye on him. In spite of the age difference, the two brothers now became close friends, and among other things developed a passion for chemistry – something of a fad in the 1820s.

In 1822, Eras went up to Cambridge to study medicine. Bored by his studies, and a lover of the good life, he became a typical Cambridge undergraduate of his day, doing the minimum amount of work and deriving the maximum amount of pleasure from his time at the university. Charles, who visited him in the summer of 1823, was introduced to this delightful way of life at an impressionably early age. About the same time, still only 14, he developed a passion for hunting, and for several years (much to his later shame) would take great delight in shooting just about anything that moved.

By 1825 his father, Dr Robert, was becoming alarmed by the way young Charles was developing. He took him out of school and made him spend the summer assisting in his medical practice, then packed him off to Edinburgh, at the age of 16, to study medicine. For the doctor, this was both a counsel of despair (he had decided that Charles would never make the grade in an academic subject like law) and the continuance of a family tradition which meant a lot to Dr Robert. But more disappointment was in store. To Charles it was all a great lark, not least because Eras, three years into his own medical studies, had decided to spend his external hospital year – his 'work experience' – in Edinburgh.

Very little work got done by either of the Darwins that academic year, but Charles did discover that he could never become a doctor. He was physically sick when asked to dissect a corpse, and fled from the room when he was supposed to be watching an operation on a child (remember, this was before the days of anaesthetics). When he returned to Edinburgh, without Eras, in 1826, he knew that his medical studies were a sham, and he began to follow geological pursuits. Geology was a new and exciting science in the 1820s (rather like cosmology in recent times), and Darwin lapped it up. In the summer of 1827 he decided that he would have to admit to his father that there was no hope of him ever becoming a doctor.

Postponing the evil day, Darwin visited London, where he met up with his cousin Harry Wedgwood, and Paris, where he enjoyed the company of Harry's sisters Fanny and Emma, before heading back to The Mount in August. Very few options were now open to Charles and his father. As Dr Robert saw it, the only hope of saving his son from a life of debauchery, squandering his share of the family fortune, was the Church, the traditional safety net for the younger sons of gentlemen. After a few months spent cramming Latin and Greek and reading religious texts, Charles passed (just) the entrance examination for Cambridge, and early in 1828 he was off to begin his studies for an ordinary BA degree, the prerequisite for taking holy orders. He certainly had no vocation for the Church, but his letters show that he quite liked the idea of being a country clergyman, which would enable him to indulge his interests in natural history and geology, and lead a comfortable life (moving on from the time of Jane Austen, we can see in the pages of Anthony Trollope's earlier works many examples of the kind of parson Darwin might have made).

Darwin worked hard enough (at last) to pass his examinations, graduating in 1831 (with a good pass, tenth out of 178), but spent more time studying botany and natural history, under the influence of the Professor of Botany, John Henslow, than on his official courses. This was nothing unusual. In those days the authorities at Cambridge took no particular interest in exactly what a young gentleman did while he was there (at least, not academically; they took a close interest in the moral side of a young gentleman's life) as long as he passed the exams. Darwin became an expert on beetles, studied geology under Adam Sedgwick, continued to enjoy the hunt, and became firm friends (nobody now can be quite sure how firm) with Fanny Owen, the daughter of one of Dr Robert's neighbours.

All good things come to an end, and in the summer of 1831 Charles went on one last (as he thought) glorious geological expedition, hammering his way across Wales, before returning to The Mount on the night of 29 August, intending, for lack of any option, to take up the career of a clergyman. Instead, he was just in time to take up an offer that would change his life completely.

The offer was waiting for him in the form of a letter from one of his Cambridge tutors, George Peacock. Peacock was passing on an invitation from Captain Francis Beaufort (he of the Beaufort scale), at the Admiralty, to join Captain Robert FitzRoy on a circumnavigation of the globe aboard the HMS *Beagle*. The main purpose of the expedition was to chart in detail large parts of the coast of South America, and would take years to complete. FitzRoy wanted a young gentleman to accompany him on the trip, a civilian social equal to diminish the isolation of command. The obvious role for such an individual on such a voyage would be as a naturalist, studying the flora and fauna of the strange lands to be visited. But, of course, the young gentleman (or his father) would be expected to pay his own way.

It is often suggested that Darwin was approached not as a naturalist, but simply as an acceptable young gentleman, who liked to hunt. This is far from being the complete truth. To be sure, he was a young gentleman, and he did like to hunt. But he had been recommended by Henslow (who had seriously considered offering himself for the role), who was well aware that Darwin was an able naturalist, had more than a smattering of geology, and would himself benefit from the experience. The one real stroke of luck for Darwin was that Henslow's first choice, another of his protégés, turned the opportunity down because he had just been appointed vicar of Bottisham. So Henslow wrote to assure Darwin that he saw him as 'amply qualified for collecting, observing, and noting any thing worthy to be noted in Natural History . . . I assure you I think you are the very man they are in search of.' Of course, Darwin would never have got the chance if he had not been a personable young gentleman from a wealthy family. But he was also a good naturalist. He was simply the best man for the job.

Darwin's father took a great deal of persuading, and the date set for departure was only a month away. Thinking that he was about to see his problematic younger son settled in a career in the Church, the last thing he wanted was for him to go gallivanting around the world. But he was persuaded to lend his reluctant support to the idea (not just moral, but financial) by his friend Josiah Wedgwood II, who helped Charles to present a clear case which answered his father's doubts and emphasized the benefits of the voyage.

Eventually, everything was agreed. Darwin met FitzRoy (who had the final say, of course, in who he wanted to accompany him on the voyage), was kitted out, and joined the *Beagle*, a three-masted vessel just 90 feet (27 m) long, carrying seventy-four men at the start of the voyage. After kicking their heels for a few weeks waiting for a fair wind, they sailed on 27 December 1831, heading first for the Canary Islands. Darwin was a few weeks short of his 23rd birthday.

The voyage would last for nearly five years, longer than even FitzRoy anticipated at the outset. While the *Beagle* was busy surveying uncharted waters, Darwin made long expeditions inland, studying not just the living fauna and flora of South America but its geology and fossil remains. He sent back to England geological samples, fossil remains and zoological specimens by the crateful, wrote long descriptive letters about what he had found to Sedgwick and Henslow, experienced a major earthquake at first hand, witnessed active volcanoes, and saw for himself how geological activity was actively raising the land from the sea. He suffered severe illness, got caught up in revolution, and visited the islands of the South Seas, including the Galapagos. The variety of living things that he saw, and the way in which life filled every available niche, set him thinking about evolution (though for a long time he would keep those thoughts to himself).

On the voyage he did not know it, but by the time the *Beagle* docked in Falmouth, on the night of 2 October 1836, Darwin was already famous in scientific circles. His reputation – as a geologist, not as any kind of biologist – was established by the rock samples and letters he had sent back from South America, which helped to establish the immense age of the Earth and the nature of the long, slow processes of geological change that have shaped it. Darwin became a Fellow of the Geological Society immediately upon his return to England; he did not bother about getting elected to the Zoological Society until 1839.

After the social whirl of activity that swept up the returned traveller, he settled down in London to work on two great projects: a book about the geology of South America, and his journal account of the voyage. In July 1837 he also started a private notebook on 'the transformation of species'. He had, of course, no need to worry about anything as mundane as earning a living. He lived off money supplied by his father, who was gratified (if somewhat baffled) to find that his problem son had made something of himself after all. He had certainly made more of himself than Eras, who had abandoned medicine and was living a carefree life in London on an allowance provided by Dr Robert: a clear indication of how Charles might have turned out without the *Beagle* experience.

To his own surprise, Charles turned out to be a gifted writer. The journal of the voyage was not published until May 1839 because it had to wait until FitzRoy finished his share of the book, describing the nautical side of the voyage. To Fitz-Roy's annoyance, Darwin's account proved so popular that it was split off to form a separate book and reprinted in August, establishing him as a scientist writing for a popular audience about subjects of widespread interest, such as the age of the Earth – the Stephen Hawking of his day. He actually made money from his writing, and moved on to a more academic book about coral reefs. In it he explained how these atolls are built up by coral growing upwards from the tops of islands which are gradually sinking beneath the waves, an explanation which still stands today. He wrote scientific papers about the new theory of ice ages. And all the while he was thinking deeply and privately about evolution. For the Charles Darwin of the early 1840s, the famous geologist, to have gone public with a theory of evolution would have seemed as bizarre to his colleagues as if a famous cosmologist in the 1980s had come up with a new theory of the origin of life. Before he could give the world the benefit of his insights into the nature of life, Darwin knew that he would have to establish a reputation as a biologist, not just a geologist.

He also had other concerns about going public with these ideas. He was now a family man, and his wife, Emma, was conventionally religious. He knew that she would be distressed both by his increasingly atheistic views and by the inevitable reaction of the Church to their publication. This was one factor (just about the only one) Darwin had not weighed in the balance when he decided to get married.

The idea of marriage, children and domestic comforts appealed to Darwin greatly after the adventure of the voyage of the *Beagle* (indeed, it had begun to appeal to him long before the voyage ended). In the summer of 1838, in his usual methodical way, he drew up a list of the advantages and disadvantages of marriage. The debit side included the risk of there being 'less money for books &c', while the credit side referred, among other things, to the desirability of a 'constant companion, (and friend in old age) who will feel interested in one, – object *to be* beloved and played with. better than a dog anyhow.' Love was not an essential prerequisite to marriage for Darwin, just as it wasn't important for many of his contemporaries. That could come later. What he wanted was somebody from a suitable background, who he knew and liked, and who preferably brought a bit of money with her.

His cousin Emma, especially given the various intermarriages between Darwins and Wedgwoods, was the obvious choice. After on extremely diffident courtship (Emma later wrote that she had 'thought we might go on in the sort of friendship we were in for years') he proposed to her on 11 November, at Maer. To the delight of both families, she accepted at once. She brought with her a dowry of £5,000 plus £400 per year, while Dr Robert bestowed £10,000 upon the happy couple to secure their future. The returned traveller, after five years away, was in no mood to delay the settling-down process, and they were married, by yet another cousin, John Wedgwood, on 29 January 1839, at St Peter's Church, Maer

– just five days after Darwin had been elected a Fellow of the Royal Society.

At the end of September 1838, Darwin began reading the *Essay on the Principle of Population* by Thomas Malthus. The key developments in Darwin's theory of evolution by natural selection seem to have taken place over about the next six months, exactly during the time of his courtship of Emma and the first few months of their marriage. By the summer of 1839 Darwin had most of the theory clear in his head, and much of it recorded, in one way or another, in his notebooks. Malthus was such a key influence that it is worth going into a little detail about just what it was he said in the *Essay*.

The *Essay* was first published, anonymously, in 1798. It dealt with how populations increase exponentially if unchecked, because (in human terms) each set of parents is capable of producing more than two offspring – a point strikingly brought home by the fecundity of both the Darwin and the Wedgwood families. At the end of the eighteenth century, the human population of North America was doubling every 25 years, a process which clearly could not be sustained indefinitely. If each pair of even the slowest-breeding animals, elephants, left just four off-spring that survived and bred in their turn, then after only 750 years each pair will have produced 19 million living descendants. In fact, over the 750 years or so leading up to Malthus's work, on average each pair of elephants alive in 1050 had just two living descendants in 1800.

Malthus realized that the potential of populations to grow wildly is kept in check by external factors such as predation, the availability of food (and, nowadays, contraception). The *majority* of individuals do not survive to reach maturity and breed. In Darwin's day, this argument was misused in some quarters to claim that the desperate way of life of the working classes in newly industrialized Britain was natural, and that it would be counterproductive to try to eliminate the causes of disease and hunger among the working population, because they would only breed until disease and hunger kept them in check once more.

Darwin saw a broader significance in Malthus's work. He looked at populations in general, not just at what was happening in the cities of Britain, and realized that in nature the individuals that did survive and breed would be the ones best adapted to their way of life – best fitted to their ecological niche, like the fit of a key in a lock (*not* the fittest in the athletic sense, though in some cases that might be one of the relevant factors). The least well adapted would be the losers in the struggle for survival, and would leave no descendants, because they would not live long enough to breed. Provided there was some variation among the individual members of a species, in each generation the fittest would tend to survive better and leave more offspring. And, crucially, Darwin realized that this struggle for existence is indeed a competition between members of the same species. It is *not* a struggle between, say, rabbits and foxes, but between rabbit and rabbit, or between fox and fox. If one rabbit can run fast enough to escape the fox, but a second rabbit cannot, it is the fast runner who survives (in this case, physical fitness does matter!). The fox doesn't mind which rabbit it eats, as long as it gets food.

By the end of 1838, even before he was married, Darwin's notebooks show that he was already drawing the comparison between this process of natural selection and the artificial selection process used by plant and animal breeders to 'improve' horses, or greyhounds, or wheat. And early in 1839 he was spelling out that there is no need for nature to 'know' which variations on the theme are good ones in order for selection to work. If it so happens that a particular bird is born with a slightly longer beak than average, and that variation helps it to get more food, then that bird will be likely to survive and breed, passing on its tendency for a slightly longer beak to the next generation. Blind chance, operating on individuals, combined with a tendency for characteristics to be inherited (but imperfectly, so that there is always the possibility of new variations arising) is all you need to explain evolution.

This was Darwin's great idea – not evolution, which was already a well-known (if not always accepted) idea, but *natural selection* – what came to be known as the 'survival of the fittest'. Critics of Darwinian ideas sometimes refer to 'the theory of evolution', but they are wrong to do so. Evolution is a fact. The full name of Darwin's theory is 'the theory of evolution by natural selection', so if you do want to use a shortened version of the name, you should refer to the theory of natural selection (which you may or may not want to criticize). Darwin had yet to give the theory a name, but by the beginning of the 1840s the clarity with which he understood what was going on is clear from his notebooks.

And yet, he sat on this idea for nearly twenty years, hugging it to himself and keeping it from the world. This didn't stop him from working on it. A thirteen-page outline of the theory, undated but apparently from 1839, survives in his papers, together with what he later referred to as a 'brief abstract', running to thirty-five pages, dated 'May & June 1842'. In the spring of 1844, he developed this version of the theory into a description about 50,000 words long, and had a fair copy of it made; he wrote a letter to Emma, to be opened in the event of his death, asking her to have this version published. But he seems at that time to have had no intention of publishing in his lifetime.

And yet ... Darwin, though not concerned about fame in his lifetime, did have the scientist's urge for his priority to be acknowledged by posterity, when and if the theory were discovered by someone else. To this end, he devised a plan so cunning that even Machiavelli would have been proud of it. During 1845, Darwin worked on a second edition of his successful journal of the *Beagle* voyage, and added new material to the existing descriptions of the living things he had seen in South America. These new passages look innocuous enough in themselves. But, as Howard Gruber pointed out in his book *Darwin on Man* (Wildwood House, London, 1974), if you compare the first and second editions of the journal you can locate all the new material, take it out of the second edition, and string it together to make a coherent 'ghost essay' which conveys almost all of Darwin's thinking about evolution at the time. It is quite clear that this material must have been written as that coherent essay, then carefully chopped up and inserted into the journal. Darwin, it seems, was torn between keeping quiet about his theory and telling it to the world.

The situation was complicated by his personal life. First, as we have seen, by his desire to avoid hurting Emma. Also, Darwin himself had been plagued by illness, off and on, ever since he returned from his travels, and he would be for the rest of his life. Whole books have been written about this illness, some suggesting that it was entirely psychosomatic, brought on by worrying about the likely public reaction to his ideas about evolution; others seeking a physical cause of one kind or another. The most likely explanation seems to us to be that it was a genuine physical ailment, very probably something he had picked up in the tropics, which was exacerbated by overwork and worry. Either way, Darwin did not see himself as the kind of robust individual that would be needed to champion these ideas publicly.

The family, too, was growing, and he had an increasing number of dependents. Charles and Emma produced ten children between 1839 and 1856 (when Emma was 48), seven of whom survived into adulthood (one, Leonard, lived until 1943). Two died in infancy; one, Darwin's favourite daughter, Annie, died in 1850, at the age of 9, plunging Darwin into a deep depression.

There was also unrest in the outside world in the 1840s and 1850s, sufficient on its own to make Darwin reluctant to poke his head above the parapet. He and Emma had settled in London after their marriage; as early as 1840 there were riots in the streets as the Chartist movement sought reform of the electoral system. The Darwins decided that London was no place to bring up their family. With financial assistance from Charles's father they purchased Down House, in the village of Down in Kent, and moved there in the middle of September 1842, a month after the army had been out on the streets of London subduing the latest violent demonstrations. A few years after the move, which made Darwin a member of the squirarchy, the spelling of the village name changed to Downe, but the name of Down House remained the same. Darwin lived there for the rest of his life, only two hours by carriage from the centre of London, but deep in the heart of the Kentish countryside. In spite of the growth of roads and spread of houses since then, the house, together with its gardens, still remain almost as he left it.

A week after the move, Emma gave birth to their third child, Mary Eleanor, who survived for only three weeks. It was a gloomy start to what turned out to be a long and happy life in Down House. William (born in 1839) and Annie (born in 1841) thrived in the country environment and, almost as if to prove the Malthusian point, within three months of losing Mary Eleanor, Emma was pregnant again.

So it was in Down House that Darwin, surrounded by a growing family and with all the domestic comforts he had craved, completed, for the benefit of posterity, his 1844 summary of the theory of evolution by natural selection, and set about making his name as a biologist so that his idea would be taken seriously. He had written a book about volcanic islands, which was published in 1844, and his long-delayed *Geological Observations on South America* was published in 1846, ten years after he had returned to England in the *Beagle*. In October that year, he wrote to Henslow:

> You cannot think how delighted I feel at having finished all my *Beagle* materials except some invertebrata; it is now ten years since my return, and your words, which I thought preposterous, are come true, that it would take twice the number of years to describe, than it took to collect and observe.

The invertebrata that Darwin referred to in that letter were some peculiar barnacles he had picked up on the shores of southern Chile in 1835. Thinking that they would make a fine study to establish his credentials in biology, and intrigued in any case by these bizarre creatures, each the size of a pinhead, that lived within the shells of other barnacles, he set to work enthusiastically.

In order to explain the place of these creatures in the barnacle family, it turned out that he had to work out the complete classification of barnacles, for the contemporary scheme was a hopeless mess. It was five years before the first of his epic volumes on barnacles appeared in 1851, produced in spite of bouts of crippling illness, Annie's death, and Darwin's increasing despair at the Herculean task he had taken on. In 1854, the final two volumes of what had turned out to be a trilogy appeared. Together, they still form the definitive work of barnacle taxonomy, and they were received with acclaim by the scientific community, earning Darwin (together with his work on coral reefs) the Royal Medal of the Royal Society. It had been far harder work than he had imagined, and taken far longer than he had expected. But by the end of the 1850s Charles Darwin was established as a biologist of the first rank, one who understood all about the relationships between species.

In the mid-1850s, Darwin still had no intention of publishing his theory of natural selection. He was 46 in 1855, a middle-aged, semi-invalid with responsibilities that extended beyond his own family. His father had died on 13 November 1848, and in the months that followed, Darwin's own illness became so severe that he resorted to the fashionable 'water cure': early-morning sessions in which he was first wrapped in blankets and heated with a spirit lamp until he was running with sweat, then plunged into ice-cold water. He also went on a special diet, without any sugar, salt, bacon or alcohol, and was allowed to work for only a few hours a day – one reason why the barnacle study took so long. Whatever the reasons, the treatment seemed to work, and Darwin stuck with it, more or less, for most of the rest of his life, though he still suffered bouts of chronic illness.

In the outside world, 1851 saw the Great Exhibition in the Crystal Palace. On 13 May, only three weeks after Annie had died, Emma gave birth to the Darwins' ninth child, Horace; so it was in July that the entire family spent a week staying with Erasmus in London and visiting the Exhibition. The visit also gave Darwin an opportunity to renew his friendship with Joseph Hooker, one of the few people to whom he had confided his theory of natural selection. Hooker was a young botanist (born in 1817), the son of the Director of the Royal Botanic Gardens at Kew, whom Darwin had met at the end of 1843, shortly after Hooker had returned from an expedition to the Antarctic as naturalist with a naval expedition. He became one of Darwin's closest scientific confidants, someone ideas could be bounced off. It had been a great blow to Darwin when Hooker went off on another expedition, to the Himalayas in 1847, and Darwin was delighted to have him back.

By now the Darwins were extremely wealthy (Charles had inherited £40,000 from his father to add to their steadily accumulating fortune) with investments in industry, railways and land. But the need to manage these investments was a constant source of worry to Darwin, who also sought to be a model landlord, looking after his tenant farmers, and diligently carried out his squirely duties around Downe, helping the poor and needy. These worries were exacerbated by the Crimean War, which started in 1854 and which led to the fall of the government and economic uncertainty in Britain. On top of all this, in December 1856, at the age of 48, Emma gave birth to their last child, Charles Waring. He suffered from what seems to have been Down's syndrome, and died less than two years later, in June 1858, of scarlet fever. The death came when Darwin was in the midst of a frantic burst of work triggered by the discovery that, while he had been prevaricating and hiding ghost essays among his papers for later generations to discover, someone else had come up with the idea of natural selection.

Darwin had not been completely scientifically inactive in the mid-1850s. For public consumption, he had carried out work on how seeds could spread around the world: testing how long they could survive in salt water and still sprout, showing how they could be spread through bird droppings, and even floating a dead pigeon in salt water for 30 days before recovering seeds from its crop and growing plants from them. These were important studies, which nobody before had bothered to carry out, and emphasize the way in which Darwin was a 'hands-on' naturalist, not an abstract armchair thinker. When he wanted to know more about artificial selection, for example, he joined pigeon-breeding societies, and learned first hand about how characteristics are passed on from one generation to the next, and how different variations on the same basic body plan can be selected by the breeder. He really did understand how living things behaved from direct, practical experience.

Privately, at this time Darwin was working on a book-length version of his theory of evolution by natural selection. He was coming round to the view that he might publish this eventually, when he was so old that it could do him no harm, and from time to time mentioned his progress with the epic work to the few people in the know. These included Hooker and the geologist Charles Lyell, who had been a major influence on Darwin at the time of the voyage on the *Beagle*. (Lyell, born in 1797, wrote an epic three-volume work of his own, the *Principles of Geology*. Darwin took volume one with him on the *Beagle*; volume two caught up with him during the voyage; volume three was waiting for him when he returned home.)

Also in Darwin's circle at this time was Thomas Henry Huxley, whom he had met at a meeting of the Geological Society in April 1853. Another naturalist who had cut his teeth on a long voyage with a naval surveying ship, Huxley was an impoverished young firebrand (born in 1825). He raged at the way science in England seemed to be the preserve of gentlemen of independent means (like Darwin), but was shrewd enough to appreciate the need for patronage from the likes of Darwin. He quickly came to see in Darwin a kindred revolutionary spirit, at least in scientific terms, for all his wealthy background. By the mid-1850s Huxley and Hooker were also close friends, members of a young generation that would indeed make science a real profession, instead of the plaything of gentleman amateurs.

Darwin's own state of mind about his evolutionary ideas can be gleaned from a letter he wrote to Lyell in November 1856:

> I am working very steadily at my big book; I have found it quite impossible to publish any preliminary essay or sketch; but am doing my work as completely as present materials allow without waiting to perfect them.

The key word is 'completely'. Darwin wanted to dot every *i* and cross every *t* before considering publishing his ideas. He wanted to consider every possible argument against his theory, and find a counter-argument that could appear in his epic book, leaving would-be critics without a leg to stand on. He was *not* going to go off at half-cock, publishing a version of the theory which would be correct in broad outline, but which might contain errors of detail with which opponents could damn him and the theory.

Had Darwin had his way, the theory of natural selection would eventually have been published (perhaps not until after his death) as a massive three-volume tome, densely packed with examples and arguments, and destined to be read only by a few scientists. That he was forced by circumstances into rushing into print what he considered to be merely an outline of the theory, an abstract written in beautifully clear, accessible language, a book which became an instant best-seller and has remained in print ever since, is one of the greatest strokes of fortune there has ever been for science. It happened like this.

Alfred Russel Wallace was a naturalist who, in 1858, was based in the Far East. He had been born in 1823, the eighth of nine children, and was obliged to leave school at 13 to earn a living. After an eventful early life, including work as a surveyor on the canals and as a schoolteacher, Wallace became a kind of freelance naturalist, supporting himself by selling specimens from around the world which he collected and sent back to England to just those wealthy gentleman amateurs that Huxley (in principle, although, as we have seen, not always in practice) despised (including Darwin). Wallace came to the theory of natural selection by exactly the same route that Darwin had, through direct experience of the fertile workings of nature in the tropics, and even through combining these direct observations with a reading of Malthus's *Essay*. Wallace corresponded with Darwin, initially mainly about specimens, not knowing that Darwin had already come up with the theory of natural selection. In September 1855, Wallace published a paper in the journal *Annals and Magazine of Natural History* which presented some of the evidence for evolution, but without introducing the idea of natural selection.

Some of Darwin's friends were sufficiently alarmed by the similarities of parts of this paper to aspects of Darwin's own work that they urged him to publish something himself. Lyell, who was by no means convinced that Darwin was right, but had the true scientist's instinct for priority, wrote to him on 1 May 1856: 'I wish you would publish some small fragment of your data, *pigeons* if you please and so out with the theory and let it take date and be cited and understood.'

Darwin, though, wasn't worried about Wallace, who he felt, from reading Wallace's paper, had some good ideas but still favoured the view that species were created by God. Evolution, after all, was nothing new, and there was no hint in Wallace's paper of natural selection. Under pressure from his friends, Darwin began to consider writing a shorter account of his own work for publication, but as late as May 1857 he was writing to Wallace in sublime ignorance of how far Wallace had gone, sending him a carefully worded 'keep off' notice, intended to establish that Darwin was the leader in these matters. Although he handed Wallace the sop of commenting 'we have thought much alike and to a certain extent have come to similar conclusions', he went on:

> This summer will make the 20th year (!) since I opened my first note-book, on the question how and in what way do species and varieties differ from each other. I am now preparing my work for publication, but I find the subject so very large, that though I have written many chapters, I do not suppose I shall go to press for two years.

Wallace didn't take the hint. If anything, what he took as Darwin's endorsement of his ideas encouraged him to write up a full account of them, including natural selection (though he did not give it that name). The result, a paper entitled 'On the tendency of varieties to depart indefinitely from the original type', was mailed to Darwin with a covering letter asking him to show it to Lyell. It landed on Darwin's desk on 18 June 1858, prompting him to send the manuscript on to Lyell with a covering letter of his own:

> Your words have come true with a vengeance – that I should be forestalled . . . I never saw a more striking coincidence; if Wallace had my MS. sketch written out in 1842, he could not have made a better short abstract!

Of course, it was *not* a coincidence: it was a scientific truth, waiting to be discovered by anyone who had eyes to see. Darwin's immediate reaction was to give up any hope of establishing his priority, though he did hope that his book might still prove useful as an explanation of the application of the theory. His friends disagreed. Lyell and Hooker, scheming independently of Darwin (now distracted by illness and then, on 28 June, by the death of baby Charles) came up with a plan to present a 'joint paper' by Darwin and Wallace to the next meeting of the Linnaean Society, due on 1 July. Darwin's contribution was largely based on his unpublished 1844 essay (Darwin had, incidentally, been 35 in 1844, the same age Wallace was in 1858); Wallace's contribution was essentially the paper he had sent to Darwin. Wallace, on the other side of the world, wasn't even consulted, but always accepted the *fait accompli* with good grace, counting himself lucky to be mentioned in the same breath as Darwin.

After all the fuss in Darwin's inner circle, the joint paper went down like a lead balloon. Almost a year later, at a meeting held on 24 May 1859, the President of the Linnaean Society summed up the previous 12 months of the society's activities with these words:

> The year which has passed . . . has not, indeed, been marked by any of those striking discoveries which at once revolutionise, so to speak, the department of science on which they bear.

Before the end of 1859, though, the cat was well and truly out of the bag.

Urged on by Lyell, and worried that Wallace might publish a book, Darwin dropped everything in the summer of 1858 and wrote his masterpiece, lifting large chunks from the existing chapters mentioned in his letter to Wallace, revising and honing his 'abstract'. The way his mind was running, and his plans to publish a full version of the theory later, can be seen from the original title: *An Abstract of an Essay on the Origin of Species and Varieties through Natural Selection.* His publisher, John Murray, managed to persuade Darwin to reduce this to *On the Origin of Species*, but the author insisted on keeping *by Means of Natural Selection* as a kind of subtitle, and including on the title page the words *or the Preservation of Favoured Races in the Struggle for Life.* Given his head, Darwin would have put the whole theory in the title! As for being an abstract, the first edition of the book, published in November 1859, ran to more than 150,000 words.

Murray did not expect to have a best-seller on his hands. He initially planned to print only 500 copies of what he thought would be a scientific tome, but increased the run by 750 when he read the finished manuscript. All 1,250 copies were in the shops on the day of publication, and a new edition followed immediately. Darwin continued to make changes as the book went through several editions in his lifetime. Most of the changes were made in an attempt to shore up the theory against criticisms which are now known to be misguided. In particular, Darwin never knew how heredity works (DNA was not identified as the carrier of the genetic code until the 1950s), and went further and further down a blind alley trying to explain it. The result is that the definitive edition of the *Origin*, as it is always referred to, is indeed the first, and that is the one you should seek out if you want to learn about natural selection and evolution from the horse's mouth. But there's no need to look for it in antiquarian bookshops; it has been reprinted many times, and is available in a cheap Penguin edition.

The theory has two key ingredients. Individuals in one generation reproduce to make new individuals in the next generation that are similar, but not identical, to their parents. That provides variety. Then, natural selection ensures that the individuals best fitted to their environment reproduce most effectively in their turn, and leave more offspring. That's all there is to it. But the implication is that all life on Earth, including human life, has arisen in the same way from some common ancestor. So people are just one species among many, not specially chosen or created by God.

Life would never be quite the same for Darwin after the publication of the *Origin.* By and large, he kept his head down at home and left the promotion of the idea of evolution by natural selection to his supporters such as Hooker and Huxley. Ironically, Huxley's fervent enthusiasm for the idea (which led to him being described as 'Darwin's bulldog') meant that he was promoting the work of an independently wealthy gentleman scientist, to some extent at the expense of a working professional scientist, Wallace. There is no need to go into the details of the debate that raged around Darwin's (and Wallace's) theory in the 1860s. History records who won that debate and why – because the facts were, and are, on the side of natural selection.

With the great work off his chest at last, and in the safe hands of a younger generation of disciples, Darwin picked up more or less where he had left off after the rude interruption caused by the arrival of Wallace's paper in 1858. He had long been fascinated by orchids, and the way in which they are fertilized by insects that are superbly fitted by evolution for that task. This fascination led to a book, bearing the archetypal Darwin title *On the Various Contrivances by which British and Foreign Orchids are Fertilised by Insects, and on the Good Effects of Intercrossing*, published in 1862. Over the next few years he prepared another magnum opus, *The Variation of Animals and Plants under Domestication*, published early in 1868, a few weeks before Darwin's 59th birthday, which spelled out the details of artificial selection, and thereby, by analogy, provided more ammunition for the arguments in support of natural selection. He began to receive honours from abroad during the 1860s (though curiously, he was never properly honoured in England during his lifetime, except in strictly scientific circles like the Royal Society).

At home, everything had settled down nicely. The surviving children were all well, and the boys all went on to successful careers in professions as diverse as banking, the army and, in the case of Francis and George (who were both, unlike their father, knighted in recognition of their achievements), science.

There were blacker moments. In 1865, FitzRoy, always a volatile personality, who had become a Bible-thumping creationist, cut his own throat. The following year, Darwin's sister Catty died, and Susan followed a few months later. Professionally, though, a clear sign of which way the wind was blowing came in the summer of 1868, when Hooker was elected President of the British Association for the Advancement of Science. As Darwin entered his own sixties, he might have been expected to rest on his laurels in easy retirement, enjoying what family life was left to him. Instead he produced his second famous book, the long-awaited volume on humankind's (or, as he put it, 'man's') place in evolution, which he had promised in the famous sentence in the *Origin*: 'light will be thrown on the origin of man and his history'.

Incidentally, his use of the term 'man' for humankind might today be regarded as reactionary and sexist. But ponder this. Darwin was actually a far-seeing revolutionary, not a reactionary: in his use of the lower-case 'man' instead of the capitalized 'Man' in the text of his book, he was deliberately ignoring the tradition of using the capital to denote the importance of our species. Fashions change, and what matters is to see these terms in the context of their times.

The light that Darwin himself shed on the subject of the origin of our species came in 1871, in a book that was really, as its title implies, two books in one – *The Descent of Man, and Selection in Relation to Sex*. Like the *Origin*, it is usually referred to by a single word, as the *Descent*. In spite of Darwin's best efforts as a skilful writer to weave the two halves together in chapters that describe sexual selection at work in people, *Selection in Relation to Sex* really follows as an afterthought to *Variation*, and *The Descent of Man* stands alone as a summary of humankind's place in evolution. Again, this is not the place to labour the point of what, by and large, is no longer a controversial theory. But the importance of Darwin's approach to the subject, in the context of his time, is best put into perspective by a passage from an essay he wrote as early as 1839, but did not publish. He said (and note the use of the capital M in this earlier work):

> Looking at Man, as a Naturalist would at any other Mammiferous animal, it may be concluded that he has parental, conjugal and social instincts, and perhaps others.

This is the nub of what Darwin did that was new to science – *looking at Man, as a Naturalist would at any other Mammiferous animal.* No special pleading, no role for God, no suggestion that humankind is in any moral way 'superior' to other animals. With those words, Darwin became the first sociobiologist. And that is the approach which carries through into the *Descent*, thirty-two years later, and which, combined with Darwin's usual clear prose, made the book another best-seller.

> Unless we wilfully close our eyes, we may, with our present knowledge, approximately recognise our parentage; nor need we feel ashamed of it. The most humble organism is something higher than the inorganic dust under our feet; and no one with an unbiased mind can study any living creature, however humble, without being struck with enthusiasm at its marvellous structure and properties.

This was Darwin's last great book. But his enthusiasm for the marvellous structure and properties of the humblest living organism shines through in a torrent of books that now poured from the pen of the master: a monograph on climbing plants, published in 1875; *Insectivorous Plants* (1875); *The Effects of Cross and Self Fertilisation in the Vegetable Kingdom* (1876); *The Different Forms of Flowers on Plants of the Same Species* (1877); *The Power of Movement in Plants* (1880); and *The Formation of Vegetable Mould, through the Action of Worms, with Observations on Their Habits*, published in 1881, the year before he died at the age of 73. He also published five scientific papers in the last year of his life.

The biggest sensation of all Darwin's writings in the last decade of his life, though, had been *The Expression of the Emotions in Man and Animals*, which sold 5,000 copies on the day of its publication in 1872. What looks like an addendum in the *Descent* found a ready market with Victorians intrigued by the notion that people really did react to outside stimuli in the same way as the other animals, and no doubt titillated by the racy chapter on blushing, which included the following description of the 'expression of emotion' in a female patient of one Dr Browne. The lady was in bed when Dr Browne and his assistants visited her, and:

> The moment that he approached, she blushed over her cheeks and temples; and the blush spread quickly to her ears. She was much agitated and tremulous. He unfastened the collar of her chemise in order to examine the state of her lungs; and then a brilliant blush rushed over her chest, in an arched line over the upper third of each breast, and extended downwards between the breasts, nearly to the ensiform cartilage of the sternum.

But this description does not represent prurience on Darwin's part. His ability to act as a detached scientific observer had long extended even to his own most intimate emotions and instincts, and while courting Emma back in 1838 he had recorded his reactions in his notebooks, commenting on the way 'sexual desire makes saliva to flow,' and the way blushing seemed to have sexual connotations, jotting down, almost in shorthand, 'blood to surface exposed, face of man . . . bosom in woman: like erection'.

In his declining years, Darwin settled into a steady routine at Down House, rising early and taking a walk before breakfast. He worked for an hour and a half, between eight and nine-thirty, then read his letters, or rather had them read to him as he lay on a sofa, before going back to work for an hour or so and then taking another walk. After lunch he read the newspaper; then it was time for him to write his letters, using a board to support the paper as he sat in a chair by the fireplace. Another hour's work in the late afternoon, then perhaps reading a novel, before dinner, and two games of backgammon with Emma, followed by bed at half-past ten.

The settled routine was described by Francis Darwin in his *Reminiscences*, appended to the autobiography of his father, which he edited and which was first published in 1887. Francis had become very close to his parents in those years, because his wife, Amy, died in 1876, soon after giving birth to Charles and Emma Darwin's first grandchild, Bernard. Francis came back to live in

Down House, and acted as his father's assistant until Charles died.

There were family holidays, and an idyllic, sunny summer at Downe in 1881, recalled with affection by the younger Darwins years later. But Charles Darwin was beginning to outlive many of his friends and relations, and on 26 August that year he learned that Eras had died. The news plunged him into yet another bout of illness, in spite of which he made the effort to sort out his brother's affairs, arrange the funeral and look after the requirements of the will. The effort exhausted him and he never really recovered, being bedridden for much of his last winter. He died, in Emma's arms, on 19 April 1882. Better, indeed, than a dog.

Although the family intended that Charles Darwin should be buried quietly at Downe, the establishment which had neglected to offer him even a knighthood in life decided (partly thanks to lobbying by Huxley) that it was time to make amends. With the permission of Emma and William (now the head of the family), Darwin was buried in Westminster Abbey, on 26 April 1882.

Afterword

At the time Darwin died, his theory of natural selection (often known as 'Darwinism' in those days) was far from established. The debate following the publication of the *Origin* had ensured that the idea of evolution was very well established, and more widely accepted than ever before. But Darwin's theory lacked an explanation of how characteristics are passed on from one generation to the next, and why there are variations from one individual to another – even among individuals who share the same parents. In the successive revisions to the *Origin* Darwin himself backed away from natural selection and in the first edition of the *Descent* he wrote that, 'in the earlier editions of my "Origin of Species" I probably attributed too much to the action of natural selection or the survival of the fittest.'

As well as the lack of a mechanism to explain the variation that is a key to Darwinism, Darwin was in retreat because of the enormous timescale required for evolution by natural selection to do its work. In the second half of the nineteenth century, the physicists and astronomers believed that the Sun could not have been hot for more than a few million years. There were good scientific reasons for them to think this. No processes known to physics at that time could have supplied the energy required to keep the Sun shining for the hundreds or thousands of millions of years required for the variety of forms of life on Earth to have evolved through a series of tiny steps.

We shall not attempt to explain the ways in which Darwin and others sought to find a way to speed up evolution, or to explain variation (except to note that this timescale problem is the main reason why Lamarckism looked attractive to many biologists at that time). It is enough to know that the information available to physicists in the nineteenth century was incomplete. They could not know about nuclear fusion, the conversion of hydrogen into helium, and how it could release energy in quantities amply sufficient to keep the Sun shining for thousands of millions of years. The energy released by nuclear fusion comes from the conversion of mass into energy, in line with Albert Einstein's famous equation $E = mc^2$; so there is an intimate link between Darwinism and the special theory of relativity.

In the twentieth century, astrophysicists established that the long timescale required for evolution by natural selection (and, incidentally, by geology) is not problematic. The Sun is now known to have been shining, essentially unchanged, for some 4½ billion years, ample time for evolution to do its work.

As for heredity, although the first steps towards an understanding of this process had been taken by an obscure Moravian monk, Gregor Mendel, in the 1850s (just when Darwin was working on the *Origin*), his ideas were not noticed at the time. They were rediscovered only when other researchers in the mainstream of biology carried out similar studies at the end of the nineteenth century, not long after Darwin died. It was, again, only in the twentieth century that biologists developed an understanding of genetics, and how characteristics are inherited by offspring from their parents.

By the 1940s, Darwinism (in its original form, as spelled out in the first edition of the *Origin*) had become firmly established as the mainstay of evolutionary theory, with the addition of one key factor – the discovery that tiny random changes (mutations) can occur in the genetic material (now known to be DNA). Such changes can cause the production of an individual having at least one characteristic which differs from that in the individual's parents.

These are not the mutations of science fiction, in which a normal human mother and father produce a monstrous offspring with supernatural powers, or in which a lion gives birth to a sheep. Instead, they are subtle variations in which, for example, a bird might have a beak a tiny bit longer than any bird of that species has ever had before. *If* that tiny bit longer beak confers an advantage, enabling that particular bird to find food more easily, it will leave many offspring in its turn, some of whom will inherit the genes responsible for the longer beak. Over many generations, such processes can create new varieties of birds, some with longer beaks and some with shorter beaks.

One recent example of the whole process at work comes from the University of California at Davis, where Francisco Ayala kept a large population of fruit flies, all descended from a single pair. The population was divided into two, one half kept in a room at 16°C, the other in a room at 27°C. Otherwise, they were treated identically, and allowed to breed. After twelve years, breeding at a rate of about ten generations per year, the average size of the flies kept in the cooler room was 10 per cent greater than that of the flies kept in the warm room. The two populations diverged at a rate of 0.08 per cent per generation, and Darwinian evolution could be seen going on before the very eyes of the experimenters.

In the 1940s, the whole package of evolutionary ideas became known as the 'new synthesis' or the 'modern synthesis'. The phrase 'new evolutionary synthesis' was coined by Julian Huxley, the grandson of Thomas Henry Huxley, in his book *Evolution: The Modern Synthesis*, published in 1942 – sixty years after Charles Darwin died, and one year before his last surviving son, Leonard, died. The publication of that book can be taken to mark the moment when Darwinism finally became established as the best explanation of how evolution works.

Just ten years later, Francis Crick and James Watson determined the structure of DNA, the molecule that carries the genetic code, and biologists were soon busy solving the code of life, and finding out how mutations work. They showed that, right down to the level of atoms and molecules, there really is no difference between the biology of human beings and that of other animals. Our genes don't just work in the same way as the genes of monkeys and apes, but in the same way as the genes of a fruit fly, or an oak tree. The correct way to understand human origins, evolution and behaviour is indeed, in Darwin's own words, by 'Looking at Man, as a Naturalist would at any other Mammiferous animal'.

A brief history of science

All science is either physics or stamp collecting.

Ernest Rutherford

c. 2000 BC	First phase of construction at Stonehenge, an early observatory.
430 BC	Democritus teaches that everything is made of atoms.
c. 330 BC	Aristotle teaches that the Universe is made of concentric spheres, centred on the Earth.
300 BC	Euclid gathers together and writes down the mathematical knowledge of his time.
265 BC	Archimedes discovers his principle of buoyancy while having a bath.
c. 235 BC	Eratosthenes of Cyrene calculates the size of the Earth with commendable accuracy.

AD 79	Pliny the Elder dies while studying an eruption of Mount Vesuvius.
400	The term 'chemistry' is used for the first time, by scholars in Alexandria.
c. 1020	Alhazen, the greatest scientist of the so-called Dark Ages, explains the workings of lenses and parabolic mirrors.
1054	Chinese astronomers observe a supernova; the remnant is visible today as the Crab Nebula.
1490	Leonardo da Vinci studies the capillary action of liquids.
1543	In his book *De revolutionibus*, Nicolaus Copernicus places the Sun, not the Earth, at the centre of the Solar System. Andreas Vesalius studies human anatomy in a scientific way.
c. 1550	The reflecting telescope, and later the refracting telescope,

	pioneered by Leonard Digges.
1572	Tycho Brahe observes a supernova.
1580	Prospero Alpini realizes that plants come in two sexes.
1596	Botanical knowledge is summarized in John Gerrard's *Herbal*.
1608	Hans Lippershey's invention of a refracting telescope is the first for which there is firm evidence.
1609–19	Johannes Kepler publishes his laws of planetary motion.
1610	Galileo Galilei observes the moons of Jupiter through a telescope.
1628	William Harvey publishes his discovery of the circulation of the blood.
1643	Mercury barometer invented by Evangelista Torricelli.

1656	Christiaan Huygens correctly identifies the rings of Saturn, and invents the pendulum clock.
1662	The law relating the pressure and volume of a gas discovered by Robert Boyle, and named after him.
1665	Robert Hooke describes living cells.
1668	A functional reflecting telescope is made by Isaac Newton, unaware of Digges's earlier work.
1673	Antony van Leeuwenhoeck reports his discoveries with the microscope to the Royal Society.
1675	Ole Roemer measures the speed of light by timing eclipses of the moons of Jupiter.
1683	Van Leeuwenhoeck observes bacteria.

1687	Publication of Newton's *Principia*, which includes his law of gravitation.
1705	Edmond Halley publishes his prediction of the return of the comet that now bears his name.
1737	Carl Linnaeus publishes his classification of plants.
1749	Georges Louis Leclerc, Comte de Buffon, defines a species in the modern sense.
1758	Halley's Comet returns, as predicted.
1760	John Michell explains earthquakes.
1772	Carl Scheele discovers oxygen; Joseph Priestley independently discovers it two years later.
1773	Pierre de Laplace begins his work on refining planetary orbits. When asked by Napoleon why there was no

	mention of God in his scheme, Laplace replied, 'I have no need of that hypothesis.'
1783	John Michell is the first person to suggest the existence of 'dark stars' – now known as black holes.
1789	Antoine Lavoisier publishes a table of thirty-one chemical elements.
1796	Edward Jenner carries out the first inoculation, against smallpox.
1798	Henry Cavendish determines the mass of the Earth.
1802	Thomas Young publishes his first paper on the wave theory of light. Jean-Baptiste Lamarck invents the term 'biology'.
1803	John Dalton proposes the atomic theory of matter.

1807	Humphrey Davy discovers sodium and potassium, and goes on to find several other elements.
1811	Amedeo Avogadro proposes the law that gases contain equal numbers of molecules under the same conditions.
1816	Augustin Fresnel develops his version of the wave theory of light.
1826	First photograph from nature obtained by Nicéphore Niépce.
1828	Friedrich Wöhler synthesizes an organic compound (urea) from inorganic ingredients.
1830	Publication of the first volume of Charles Lyell's *Principles of Geology*.
1831	Michael Faraday and Joseph Henry discover electromagnetic induction.

	Charles Darwin sets sail on the *Beagle*.
1837	Louis Agassiz coins the term 'ice age' (*die Eiszeit*).
1842	Christian Doppler describes the effect that now bears his name.
1849	Hippolyte Fizeau measures the speed of light to within 5 per cent of the modern value.
1851	Jean Foucault uses his eponymous pendulum to demonstrate the rotation of the Earth.
1857	Publication of Darwin's *Origin of Species*. Coincidentally, Gregor Mendel begins his experiments with pea breeding.
1864	James Clerk Maxwell formulates equations describing all electric and magnetic phenomena, and shows that light is an electromagnetic wave.

1868	Jules Janssen and Norman Lockyer identify helium from its lines in the Sun's spectrum.
1871	Dmitri Mendeleyev predicts that 'new' elements will be found to fit the gaps in his periodic table.
1887	Experiment carried out by Albert Michelson and Edward Morley finds no evidence for the existence of an 'aether'.
1895	X-rays discovered by Wilhelm Röntgen. Sigmund Freud begins to develop psychoanalysis.
1896	Antoine Becquerel discovers radioactivity.
1897	Electron identified by Joseph Thomson.
1898	Marie and Pierre Curie discover radium.
1900	Max Planck explains how

	electromagnetic radiation is absorbed and emitted as quanta. Various biologists rediscover Medel's principles of genetics and heredity.
1903	First powered and controlled flight in an aircraft heavier than air, by Orville Wright.
1905	Einstein's special theory of relativity published.
1908	Hermann Minkowski shows that the special theory of relativity can be elegantly explained in geometrical terms if time is the fourth dimension.
1909	First use of the word 'gene', by Wilhelm Johannsen.
1912	Discovery of cosmic rays by Victor Hess. Alfred Wegener proposes the idea of continental drift, which led in the 1960s to the theory of plate tectonics.
1913	Discovery of the ozone layer by

	Charles Fabry.
1914	Ernest Rutherford discovers the proton, a name he coins in 1919.
1915	Einstein presents his general theory of relativity to the Prussian Academy of Sciences.
1916	Karl Schwarzschild shows that the general theory of relativity predicts the existence of what are now called black holes.
1919	Arthur Eddington and others observe the bending of starlight during a total eclipse of the Sun, and so confirm the accuracy of the general theory of relativity. Rutherford splits the atom.
1923	Louis de Broglie suggests that electrons can behave as waves.
1926	Enrico Fermi and Paul Dirac discover the statistical rules which govern the behaviour of

	quantum particles such as electrons.
1927	Werner Heisenberg develops the uncertainty principle.
1928	Alexander Fleming discovers penicillin.
1929	Edwin Hubble discovers that the Universe is expanding.
1930s	Linus Pauling explains chemistry in terms of quantum physics.
1932	Neutron discovered by James Chadwick.
1937	Grote Reber builds the first radio telescope.
1942	First controlled nuclear reaction achieved by Enrico Fermi and others.
1940s	George Gamow, Ralph Alpher and Robert Herman develop the Big Bang theory of the origin of the Universe.

1948	Richard Feynman extends quantum theory by developing quantum electrodynamics.
1951	Francis Crick and James Watson work out the helix structure of DNA, using X-ray results obtained by Rosalind Franklin.
1957	Fred Hoyle, together with William Fowler and Geoffrey and Margaret Burbidge, explains how elements are synthesized inside stars. The laser is devised by Gordon Gould. Launch of first artificial satellite, *Sputnik 1*.
1960	Jacques Monod and Francis Jacob identify messenger RNA.
1961	First part of the genetic code cracked by Marshall Nirenberg.
1963	Discovery of quasars by Maarten Schmidt.

1964	W.D. Hamilton explains altruism in terms of what is now called sociobiology.
1965	Arno Penzias and Robert Wilson discover the cosmic background radiation left over from the Big Bang.
1967	Discovery of the first pulsar by Jocelyn Bell.
1979	Alan Guth starts to develop the inflationary model of the very early Universe.
1988	Scientists at Caltech discover that there is nothing in the laws of physics that forbids time travel.
1995	Top quark identified.
1996	Tentative identification of evidence of primitive life in a meteorite believed to have originated on Mars.